THE MOON AND ITS EXPLORATION

NECIA H. APFEL

FRANKLIN WATTS
NEW YORK/LONDON/TORONTO/SYDNEY/1982
A FIRST BOOK

Cover photograph of earth rising
over the moon courtesy of NASA

Interior photographs courtesy of NASA: pp. 6,
14, 30, 34, 38, 41, 42, 46 (both), 53, 55, 56, 59;
and Necia Apfel: pp. 25 (both) and 37.

The photograph on page 53 is of *Apollo 11* at lift-off.

Diagrams by Vantage Art, Inc.

Library of Congress Cataloging in Publication Data

Apfel, Necia H.
The moon and its exploration.

(A First book)
Bibliography: p.
Includes index.
Summary: Discusses the geological features,
movement, and recent exploration of Earth's closest
neighbor.
1. Moon—Juvenile literature. [1. Moon]
I. Title.
QB582.A63 523.3 81-21877
ISBN 0-531-04385-1 AACR2

Printed in the United States of America

CONTENTS

THE MOON AND ITS EXPLORATION

CHAPTER ONE

THE MOON IN HISTORY

The moon is probably the most frequently looked at object in the sky. It is usually the first thing viewed through a new telescope, and because it is so bright, it can be seen easily even during the daytime.

People throughout history have observed the moon. They have recorded its changing shapes, or *phases*; they have timed its risings and settings and its motion through the sky. The length of our month was originally based upon the time that the moon took to go through its phases. *Monday* was the "moon's day" in early calendars.

The moon has also been given many poetic names by writers and others. Some of these have entered our language. One such name was "Luna," which was the Latin word for moon. The words *lunacy* and *lunatic* come from the supposed effect of the full moon on a person's behavior. We also use the word *lunar* when we refer to the moon. However, unlike the other moons in the solar system, our moon has no name. It is simply referred to as "the moon."

Before the invention of the telescope in the early 1600s, the moon was believed to be a smooth ball, with light and dark shadings. It certainly appears that way, especially when the familiar "man-in-the-moon" face is seen.

In 1610, the famous Italian scientist Galileo Galilei turned the newly invented telescope toward the sky. The moon was one of the first objects he looked at. Imagine his surprise upon seeing deep craters, rocky crags, and mountain ranges. These were the brighter areas of the moon's surface. The darker areas looked smooth through Galileo's small telescope, and he therefore assumed that they were seas, similar to the seas here on earth. He called them *maria*, which is the Latin word for seas. (The singular of maria is *mare*.) Although all of these dark areas are still called maria, today we know that they are not bodies of water. They are huge, lava-filled craters. It is the lava that gives them their dark color.

Since the days of Galileo, we have learned much more about the moon, the closest neighbor to the earth. As telescopes got larger and more powerful, more details could be seen on the lunar surface. Then, in the middle of the twentieth century, a number of spacecraft were sent to the moon, some with astronauts aboard. The astronauts were able to walk on the surface of the moon and perform many experiments. Our knowledge of the moon has increased tremendously from all of these investigations.

CHAPTER TWO
PHASES AND MOTIONS OF THE MOON

You have probably seen the moon in all of its different phases, ranging from a thin, delicate crescent to a round, shiny saucer. Of course, the moon does not actually change shape. It appears to do so because we see different sections of its sunlit side at different times of the month. We cannot see any part of the moon unless it is illuminated by the sun. Moonlight is sunlight reflected off of the lunar surface.

As you can see in the diagram, when most of the moon's sunlit side is turned away from us, we can see only a thin crescent. The thin *crescent moon* is visible at twilight right after the sun has set. Sometimes, when the sky is very clear, the crescent moon will appear with a faint outline of the entire moon around it. Some people say that "the old moon is in the new moon's arms." We can see this faint outline because some of the sunlight hitting the earth is reflected onto the dark portion of the moon, illuminating it very dimly. Of course, this reflected light, which is called *earthshine,* is much less intense than the light that comes directly from the sun itself, but it is enough to show us the lunar outline.

About one week after the crescent moon is seen, the shape of the moon has grown to half a circle. By this time, the moon has moved one-fourth of the way in its orbit around the earth. We can see more of its sunlit side. We call the moon in this phase a *half-moon*.

Phases of the Moon

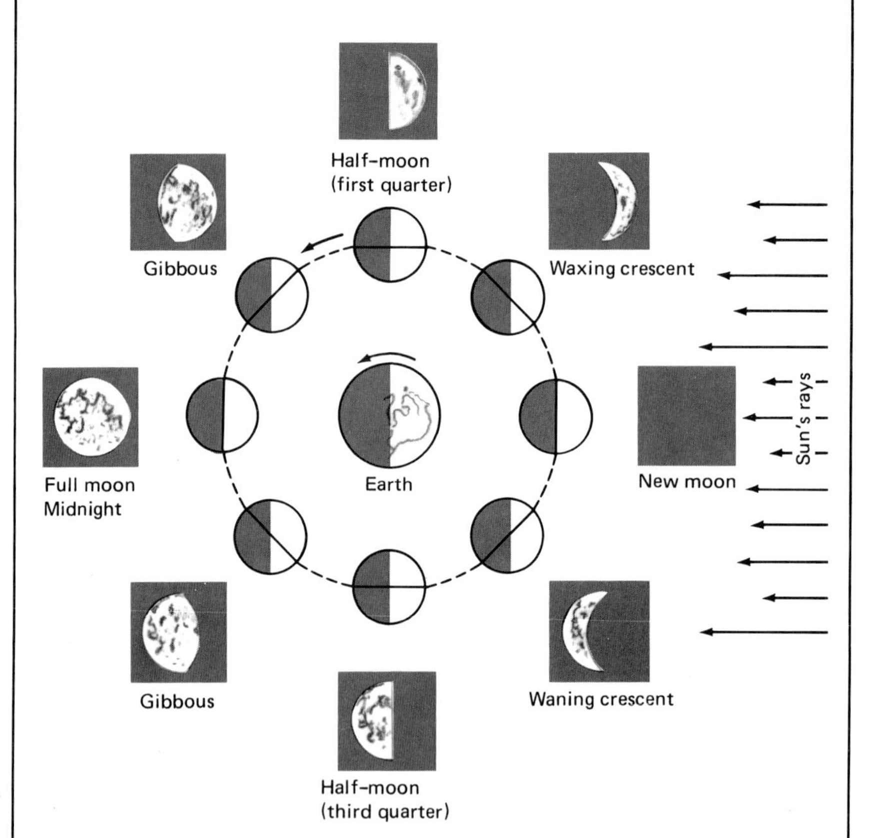

The solid lines divide the near side of the moon from the far side. Only the lit up portion of the near side can be seen from the earth.

As the moon moves farther in its orbit, more and more of its sunlit side becomes visible to us. Once it has passed the half-moon phase, it is called a *gibbous moon,* because it is becoming rounded in shape. (*Gibbous,* from the French, means "humpbacked.") Finally, when it reaches the point in its orbit where it is on the opposite side of the earth from the sun, we can see its entire sunlit side. This is a *full moon*.

From there the moon continues orbiting around the earth, appearing once again gibbous, then half, and finally crescent-shaped. We cannot see the moon when it is on the same side of the earth as the sun. This is because the side of the moon that is facing us is not lit by the sun. We call this phase the *new moon*. Within a few days, however, the slender crescent moon is seen once again. The entire cycle takes a little less than one calendar month.

Regardless of which phase the moon is in, the same side of it always faces the earth. This side is called its *near side*. The side that always faces away from the earth, and therefore can never be seen from the earth, is called the *far side*. Because the same side of the moon always faces the earth, some people mistakenly think that the moon does not rotate, that is, turn on its axis. However, as you will note in the diagram, as the moon revolves around the earth, different parts of its surface are illuminated by the sun at different times. For example, when the moon is new, the near side receives no sunlight. That is why we cannot see the moon in this phase. It is illuminated only on its far side. At full moon, the reverse is true; the sun shines only on the near side, leaving the far side in darkness. At half-moon, only half of the near side (and half of the far side) is in sunlight. The other halves of both the near and far sides are dark.

If you were viewing the moon from the sun's position, you would be able to see all sides of it. It would be seen turning around and around as it went around the earth. Every part of the surface would experience both day and night throughout each revolution around the earth. The moon rotates on its axis in the same amount of time that it takes to make one journey around the earth—approximately one month.

If you were living on the moon, you would see the earth go through the same phases as the moon does each month. Different amounts of the sunlit side of the earth would be visible at different points in the lunar orbit. The only difference between the phases of the moon and the phases of the earth is that they are exactly opposite each other. When the moon is seen as full from the earth, the earth is seen as a "new earth" from the moon. When the moon is new, the earth is a "full earth."

Because the crescent-shaped moon is on the same side of the earth as the sun, it rises shortly after the sun. It is very close to the sun in the sky during the daytime, and therefore the sun's brilliance blocks out its light at this time. We cannot see it at all. Only when the sun finally sets below the horizon can we see the slender crescent low in the western sky. Let us remember, however, that the rising and setting of the sun and moon are due to the rotation of the earth on its axis. It is because the earth is turning that the sun, moon, and other celestial objects appear to rise and set.

Meanwhile, the moon is moving around the earth, and therefore it does not remain on the same side of the earth as the sun. This means that the moon does not continue to rise and set with the sun. In order for you to see the moon rise, you must wait later and later each day until the earth has had time to turn farther around on its axis. For this reason, as the moon revolves around the earth, it rises later and later and is seen in the sky as being farther and farther away from the sun. By the time the moon is at its half-moon phase, it is rising around noon and setting around midnight. The half-moon is usually visible in the middle of the afternoon and makes a splendid sight once the sun has set.

The full moon rises around sunset and does not set until the following dawn. It is so bright that it blocks out the light of

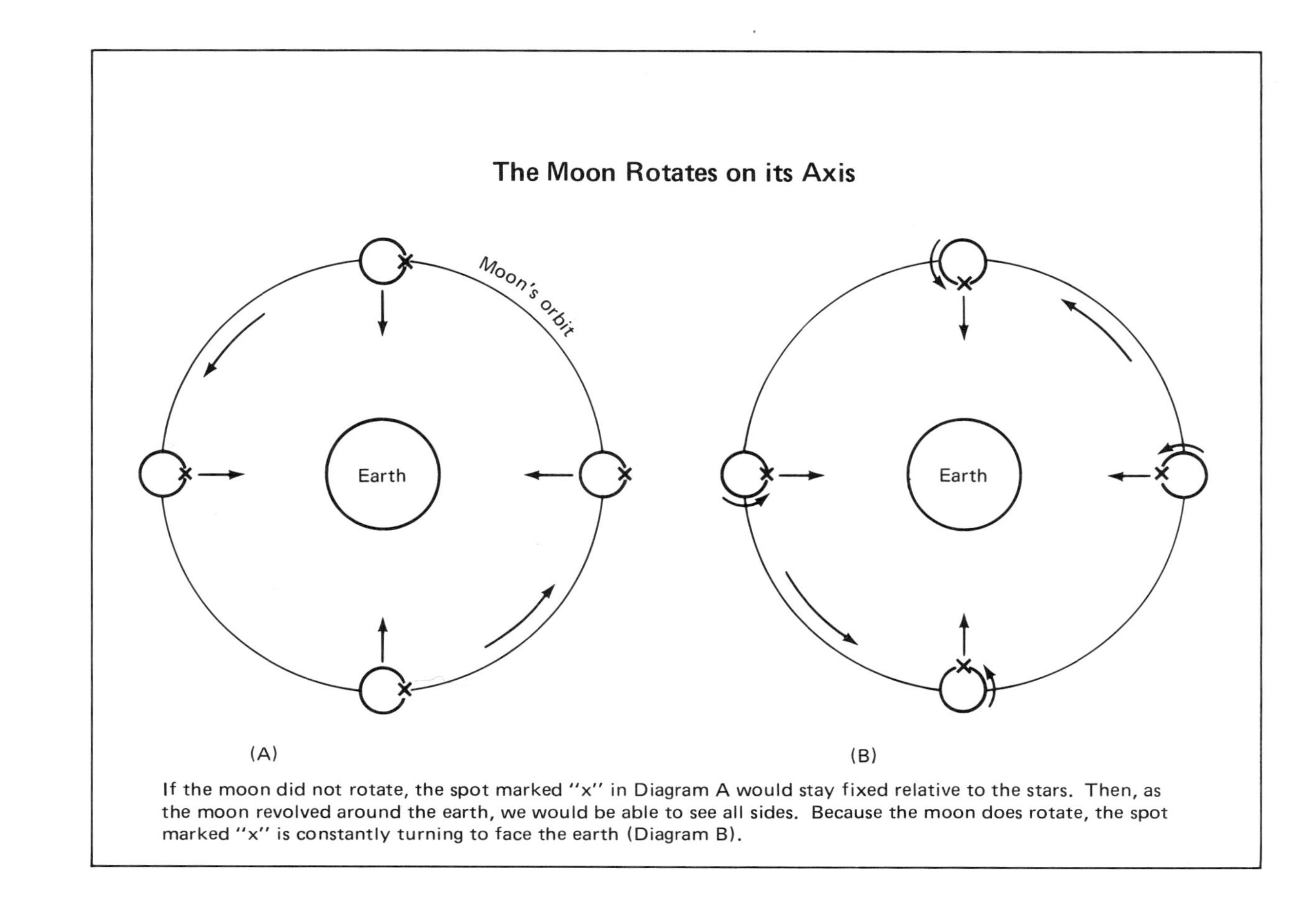

If the moon did not rotate, the spot marked "x" in Diagram A would stay fixed relative to the stars. Then, as the moon revolved around the earth, we would be able to see all sides. Because the moon does rotate, the spot marked "x" is constantly turning to face the earth (Diagram B).

The earth as seen from the moon.

The Zodiac

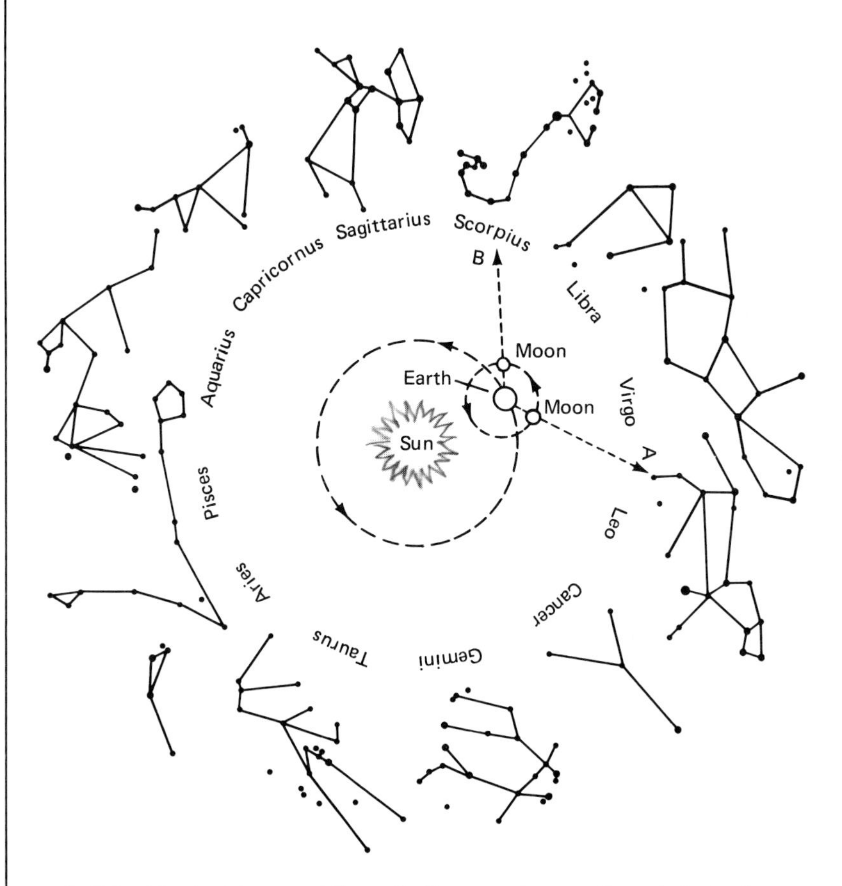

As the moon revolves around the earth, it appears to move through the twelve constellations of the zodiac. At Point A, it appears in the constellation of Leo; at B, it is in Scorpius.

all but the brightest stars. Because farmers in the past were able to continue to harvest their fall crops by the light of the full moon, the full moon that occurred nearest the first day of fall was called the "harvest moon."

Because you cannot see the moon rise in the week following the new-moon phase (it is lost in the glare of the rising sun), you will not notice that it is rising later and later each day. But if you watch the moon every evening that it is visible, you will discover that it sets later and later each night. In fact, it rises and sets an average of fifty minutes later each day.

Because of this, the moon not only changes its position in the sky relative to the sun, it also changes its position relative to the stars. Throughout the month, it follows a set path across the entire sky, passing through a group of twelve constellations called the *zodiac.* The sun and all the planets also move through these constellations, changing their positions in the sky relative to the stars, but none do so as rapidly as the moon. The sun, for example, takes a whole year to cover the entire band of zodiac constellations, moving through one each month. The moon makes this same motion in the sky every single month. If you are familiar with the constellations, you can record which one the moon is in each night.

CHAPTER THREE

THE DISTANCE TO THE MOON

The moon is the closest body to the earth. Except for the sun, it appears as the largest object in the sky. Actually, the sun and moon appear to be about the same size. This is because, although the sun is about 400 times bigger than the moon, it is also approximately 400 times farther away.

Two thousand years ago, astronomers were able to roughly calculate the distance to the moon by using simple geometry. More accurate measurements were made in the centuries that followed, but it was not until the development of radar in the middle of the twentieth century that an entirely new and very precise method of measuring the moon's distance was developed.

Project Diana in 1946 was the first to use this new technique. Astronomers measured the time it took for radar waves to travel to the moon and back—about 2½ seconds. They knew that radar waves travel at the speed of light (about 186,300 miles, or 298,080 km, per second) and could therefore easily determine the moon's distance using simple mathematics.

However, the most precise method for measuring the moon's distance was made possible by the *Apollo* astronauts who traveled to the moon in spacecraft during the late 1960s and early 1970s. While on the moon, the astronauts placed special mirrors on the lunar surface so that laser beams from the earth could be reflected off of them. Laser light stays in a very narrow, pencil-like beam instead of spreading out like radar or ordinary light. It could therefore be pinpointed to specific locations on the moon and the earth. So precise was this method of measuring that the distances between points on the earth and points on the moon could be determined to within inches.

We now know that from the center of the earth to the center of the moon, the average distance is 238,857 miles (382,171 km). We usually round this off to 240,000 miles (384,000 km). We use an average because distances to the moon vary slightly throughout the month. The lunar orbit is elliptical (oval-shaped) rather than being a perfect circle, and therefore the earth-moon distance is not always the same.

This variation in the earth-moon distance is not noticeable to the naked eye, although if measurements were made of how big the moon appears to us, it would be found to appear somewhat bigger when it is closer to us than when it is farther away.

However, when the full moon is rising or setting, it sometimes looks about two or three times larger than normal. This has nothing to do with its distance from us. If you took a motion picture of the moon as it was rising, you would notice no change at all in its size. On the film, the moon does not appear any bigger when it is near the horizon than when it is high in the sky. There have been many explanations given for this phenomenon, but none has proved satisfactory. It is an unexplained optical illusion.

CHAPTER FOUR
THE DOUBLE PLANET

The earth and moon have often been called the "double planet." Although our moon is not the largest moon in the solar system, it is the largest in relation to the size of its mother planet, the earth. The moon's diameter is one-fourth that of the earth's. No other planet has a moon that is even one-tenth its size. The other large moons in the solar system revolve around giant planets, such as Jupiter and Saturn, which are many times bigger than the earth.

Because of the relatively large size of the moon compared to the earth, the moon's effect upon the earth is far greater than that of any other moon on its respective planet. One major effect the moon has on the earth is the tides. It is the moon's gravitational pull on the earth that causes the earth's waters to be drawn upward. They appear to rise and then fall as the moon passes overhead in its orbit around the earth.

You can notice the change in water level at the seashore during the course of a day. The edge of the water moves farther inland and then retreats back toward the sea. High tide is followed by low tide every six hours. There are two high tides and two low tides each day at any given place on earth.

Although the moon's gravitation mainly affects the waters of the earth, it also has some effect on the land. For example, the North American continent may rise as much as 6 inches (15 cm) when the moon is overhead.

The sun also produces tides, but it has much less effect than the moon. This is because, although the sun is so much bigger and more massive than the moon, it is also much farther away. However, when the sun and moon are on the same side or on opposite sides of the earth, as they are at the full- or new-moon phases, their combined pull on the earth creates the highest tides of the month. These tides are called *spring tides*, perhaps because the waters rise, or "spring up," so much higher than usual. When the moon is at its half-moon phase, it is at right angles to the sun as seen from the earth. At that time, the pull of the sun and the moon upon the waters of the earth counteract each other, and the water rises much less at high tide. These tides are called *neap tides*.

The constant flow of tidal waters in and out over the ocean floor creates a friction that is very gradually slowing down the rotation of the earth. This effect was first discovered when historical records of ancient eclipses (we will find out about these in the next chapter) were compared with modern, computerized timings of these same events. It was found that the computer time and the actual solar time recorded 2,000 years ago differed by about three hours. This indicated that our days are very gradually becoming longer—by about 0.0016 seconds per century!

As the earth's rotation slows down, changes in the earth-moon system must take place in order to keep everything in balance. Very slowly, the moon is moving away from the earth and is taking longer to make each revolution. The month, too, is thus becoming longer, although at a slower rate than that at which the day is lengthening. Eventually, one rotation of the earth will take the same amount of time as one lunar revolution —47 of our current days. When this happens, the earth and moon will both keep the same sides continually toward each other, creating other effects here on earth. None of these changes will be obvious to us here on earth for many hundreds of thousands of years.

CHAPTER FIVE
ECLIPSES

There are two times during every orbit of the moon around the earth when the earth, moon, and sun are in line with each other. This alignment occurs when the moon is in either its new or full phase. As we have seen, it is at such times that high tides occur. Far more exciting phenomena also occur at these times, although not as frequently as the spring tides. These are the *solar* and *lunar eclipses.* The shadow of the earth or the moon falls upon the other body, temporarily blocking out the sunlight.

We do not experience eclipses at every full or new moon because most of the time, the moon's path takes it slightly above or below the line between the earth and the sun. There are only two to seven eclipses every year, and most of them are not very spectacular. In fact, unless you are notified beforehand that such an event will occur, you usually will not notice that anything is happening. Only when a *total solar eclipse* occurs is the event truly dramatic.

At such times, the moon passes directly between the earth and the sun, casting its shadow upon a small area of the earth. As the earth rotates, this shadow moves across the land, making an *eclipse band* about 1,000 miles (1,600 km) long and less than 200 miles (320 km) wide. Only the people in this narrow band experience a total solar eclipse.

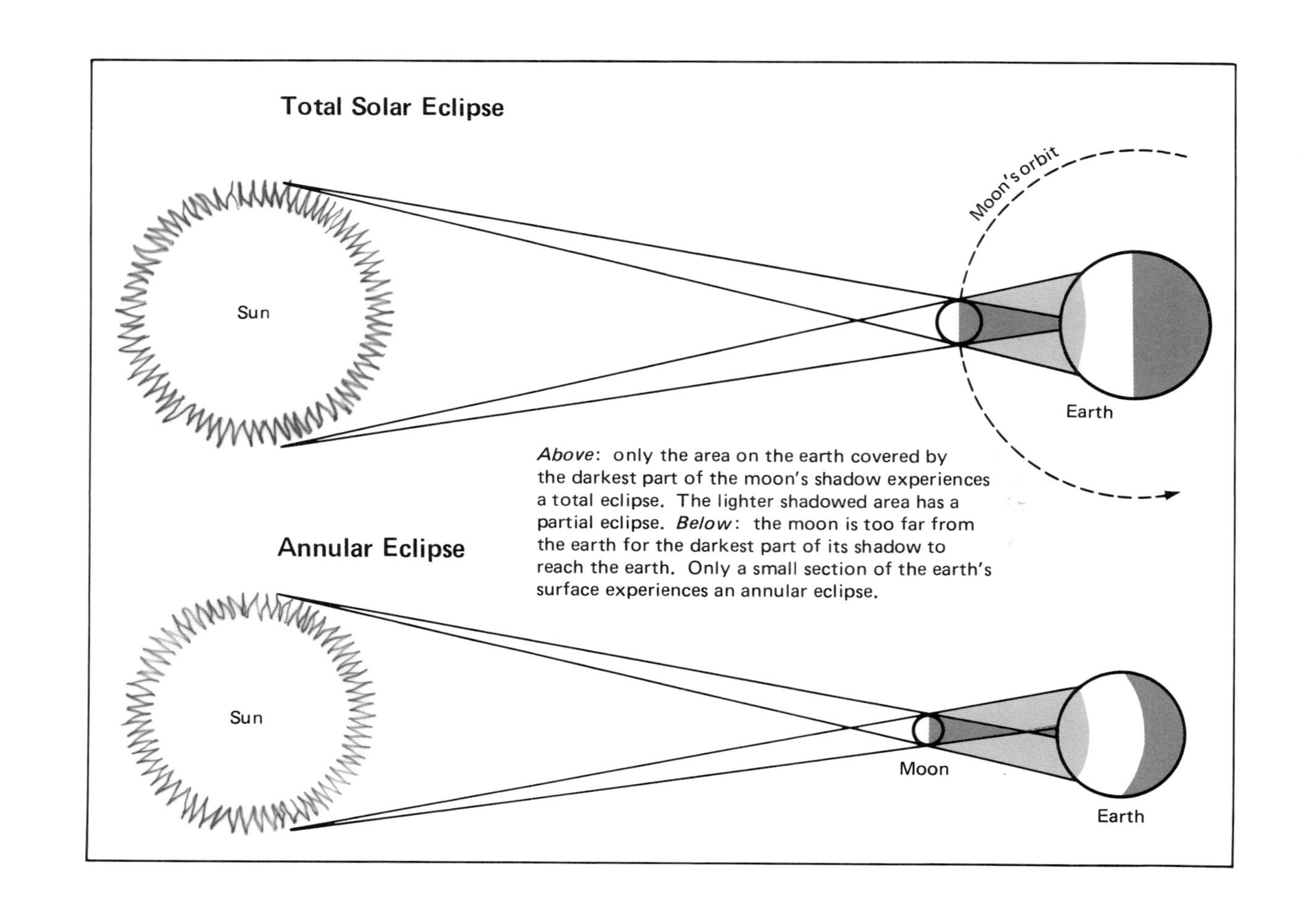

Above: only the area on the earth covered by the darkest part of the moon's shadow experiences a total eclipse. The lighter shadowed area has a partial eclipse. *Below*: the moon is too far from the earth for the darkest part of its shadow to reach the earth. Only a small section of the earth's surface experiences an annular eclipse.

For a very few minutes they cannot see the sun. All that is visible where the sun normally would be is the huge solar corona, which is the outermost part of the sun's atmosphere. The corona is so faint that it is not visible at all when the sun is shining brightly.

During the total solar eclipse, it gets so dark on earth that stars and planets can be seen in the sky. The birds, thinking that night has come early, go home to roost; cows in their pastures start returning to the barn.

Just before the moon moves directly in front of the sun and completely covers its disk, the last bright rays of sunlight can be seen shining through the mountain peaks on the moon. This produces what is known as the *Diamond Ring effect*.

Throughout history, a solar eclipse has been a fearful experience, especially in the days when its cause was not understood. At such times, many thought that the end of the world was at hand. The early Chinese thought that a dragon was swallowing the sun, and they made a great commotion in an effort to induce the dragon to release the sun. Such noisemaking during eclipses was common to many early cultures, and even occurs presently among some tribes. Naturally, the drumbeating and cymbal clanging always seemed to be a complete success. The sun always returned, shining as brightly as ever.

The areas bordering the narrow eclipse band experience only *partial eclipses*. They are not directly under the darkest part of the moon's shadow. They see the moon only covering a portion of the sun. The sun's light is somewhat diminished, but it is still much too bright to be looked at directly. This is true even if only a tiny sliver of the sun's disk is visible. Incidentally, you must *never* look directly at the sun during an eclipse (or at any other time) because its powerful rays can blind you. Use what is known as an *eclipse box* or similar device to view the sun indirectly on a piece of white paper. Smoked glass or other homemade filters are not safe.

Because the lunar orbit is an ellipse (oval-shaped) rather than a circle, there are times when the moon is too far from the earth to completely cover the sun's disk during an eclipse. We then experience an *annular eclipse* of the sun. During this kind of an eclipse, only the central portion of the sun's disk is covered by the moon. The outer rim still shines very brightly—much too brightly to be looked at directly. Annular eclipses would probably not even be noticed by anyone other than astronomers studying the phenomenon.

At the end of this chapter, there is a list of future total solar eclipses and where they can be viewed (using the proper protective devices, of course). You will note that there are no more to be seen in the continental United States until the next century. Actually, we are very lucky to be living at this time in the evolution of the earth-moon system, when eclipses are still occurring. Because the moon is very gradually moving farther away from the earth, it will eventually be too far for its shadow to ever reach the earth's surface. All that we will have then will be annular eclipses. Today, about 37 percent of all solar eclipses are annular, but in a few million years, all solar eclipses will be that way.

A lunar eclipse occurs when the earth comes directly between the moon and the sun. Although not nearly as dra-

Left: *a total solar eclipse. Only the solar corona is visible when the moon covers the entire solar disk.* Right: *the Diamond Ring effect*

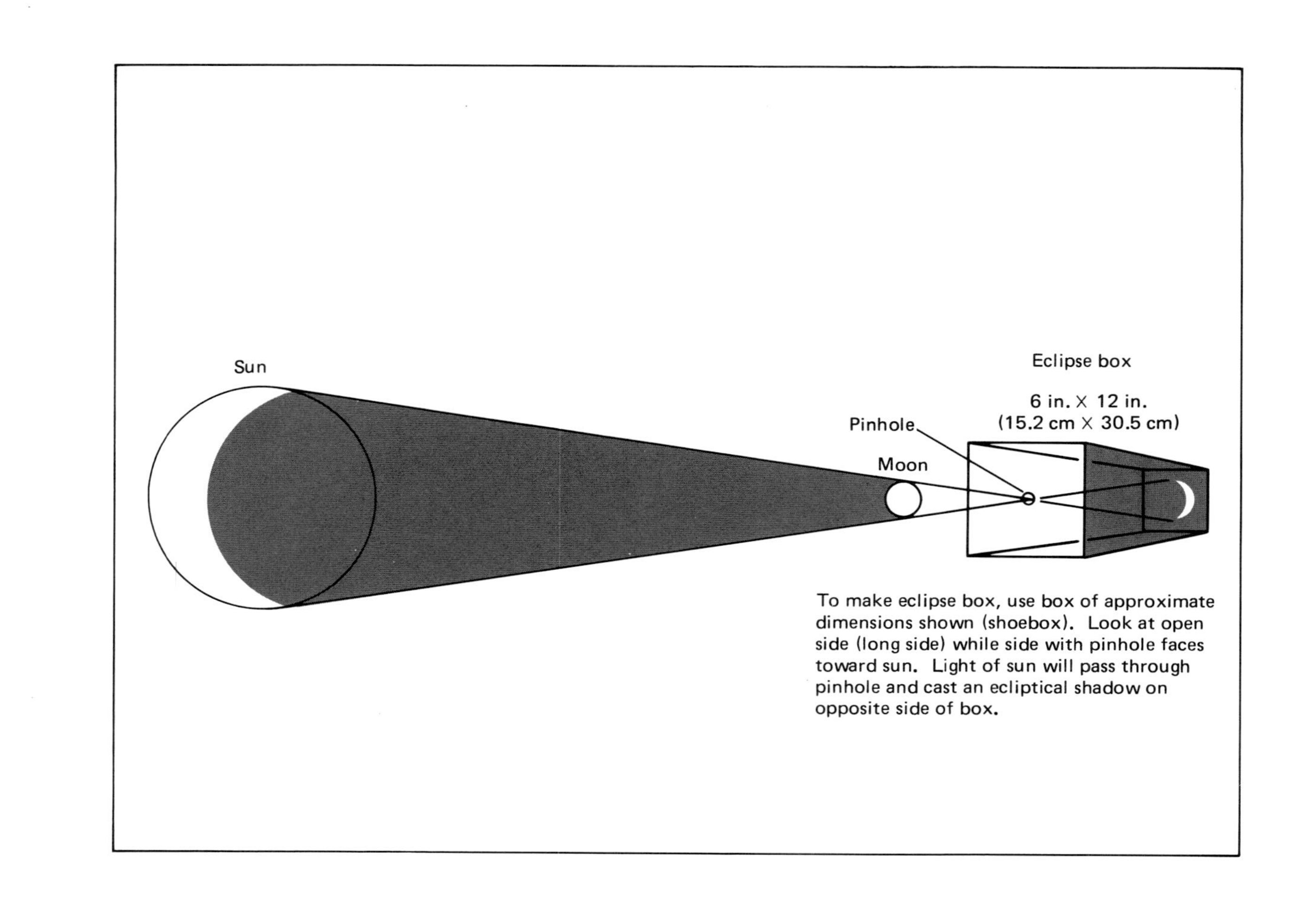
Sun
Eclipse box
6 in. × 12 in.
(15.2 cm × 30.5 cm)
Pinhole
Moon
To make eclipse box, use box of approximate dimensions shown (shoebox). Look at open side (long side) while side with pinhole faces toward sun. Light of sun will pass through pinhole and cast an ecliptical shadow on opposite side of box.

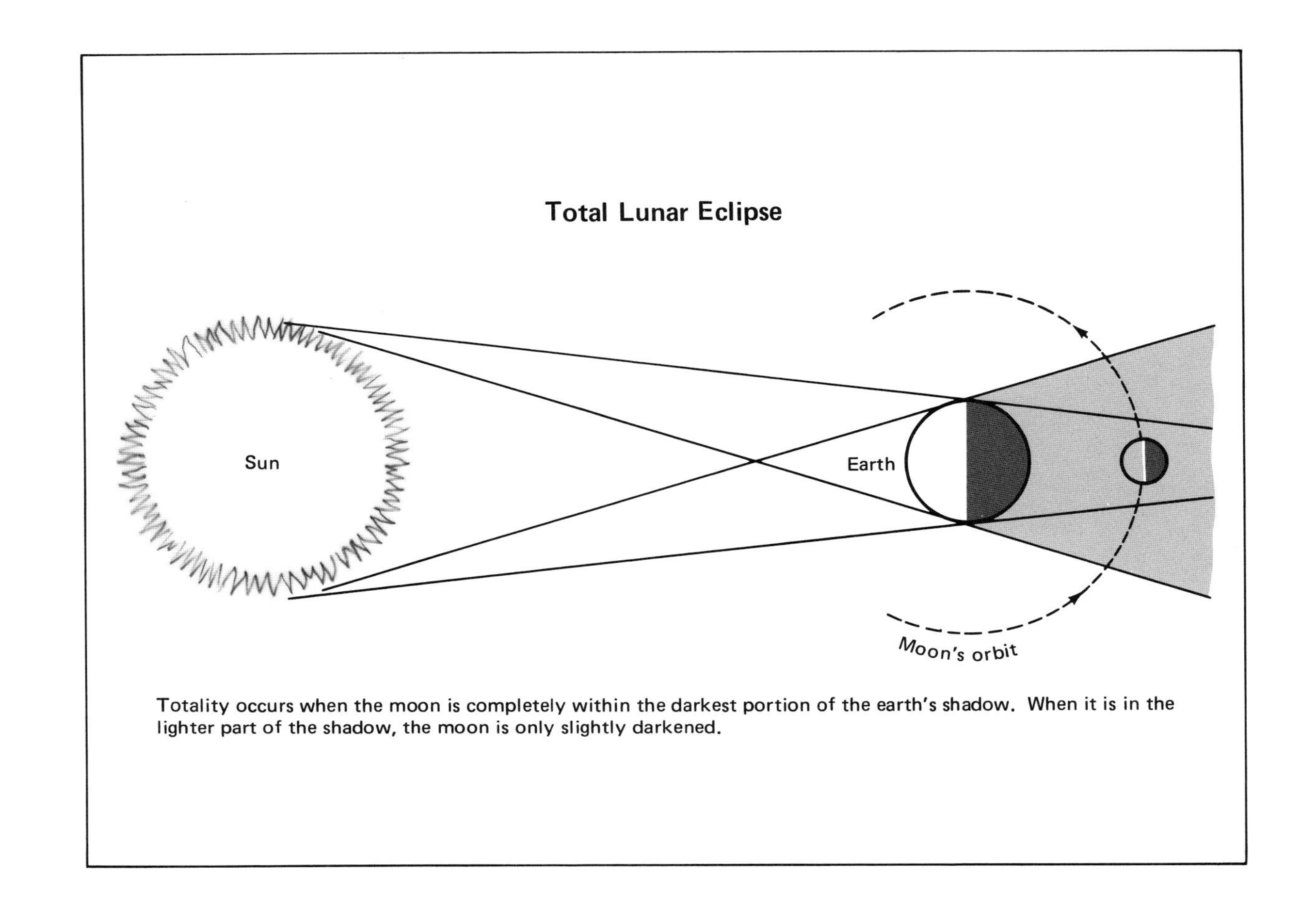

Totality occurs when the moon is completely within the darkest portion of the earth's shadow. When it is in the lighter part of the shadow, the moon is only slightly darkened.

matic as a solar eclipse, it is more easily and safely viewed and can be seen by many more people. As the moon moves through the earth's shadow, the eclipse can be seen by all the observers on the hemisphere facing the moon. A lunar eclipse lasts several hours.

You might expect the moon to be invisible during a lunar eclipse, since no sunlight is shining on it. But actually, a small amount of light does reach the moon's surface. This is the light from the sun that passes through the earth's atmosphere and is slightly bent in its path, so that it falls upon the moon, very dimly illuminating it. The eclipsed moon appears as a ruddy-colored ball hanging in the sky. It has the same dark and light features but now in tones of red. Sometimes, if the earth's atmosphere is unusually dusty (such as from a forest fire or volcanic eruption), the eclipsed moon will appear quite dark.

FUTURE TOTAL SOLAR ECLIPSES

Date	*Where visible*
June 11, 1983	Indonesia
Nov. 22, 1984	Indonesia, South America
March 29, 1987	Central Africa
March 18, 1988	Philippines, Indonesia
July 22, 1990	Finland, Arctic Regions
July 11, 1991	Hawaii, Central America, Brazil
June 30, 1992	South Atlantic
Nov. 3, 1994	South America
Oct. 24, 1995	South Asia
March 9, 1997	Siberia, Arctic
Feb. 26, 1998	Central America
Aug. 11, 1999	Central Europe, Central Asia

CHAPTER SIX

THE MOON THROUGH A TELESCOPE

When you look at the moon through a telescope, you don't see the man-in-the-moon image. One reason is that everything seen through an astronomical telescope appears upside down. This is the effect of the lenses or mirrors in the telescope, redirecting the light rays that enter it to the point where they can be viewed or photographed. Since there is no up or down in space, this upside-down view of celestial objects presents no problem to astronomers.

However, when photographs of the moon are shown in books or magazines, it is the usual custom to show the moon as it appears to the naked eye—that is, right side up. This is what has been done with the photograph on p. 6. If you look at this photograph from some distance away, you will begin to see the man-in-the-moon. A closer look will reveal what causes this "face" to be seen by the unaided eye whenever the moon is full.

The "eyes," "nose," and "mouth" are large maria. These dark areas look relatively smooth through a small telescope, especially when compared with the rest of the terrain, which consists of rough, crater-covered mountains and hills. Because the sunlight is shining directly down upon the lunar surface at the full-moon phase, it is not easy to see most of the craters and mountains clearly at that time. They cast no shadows and so do not look very deep or ragged.

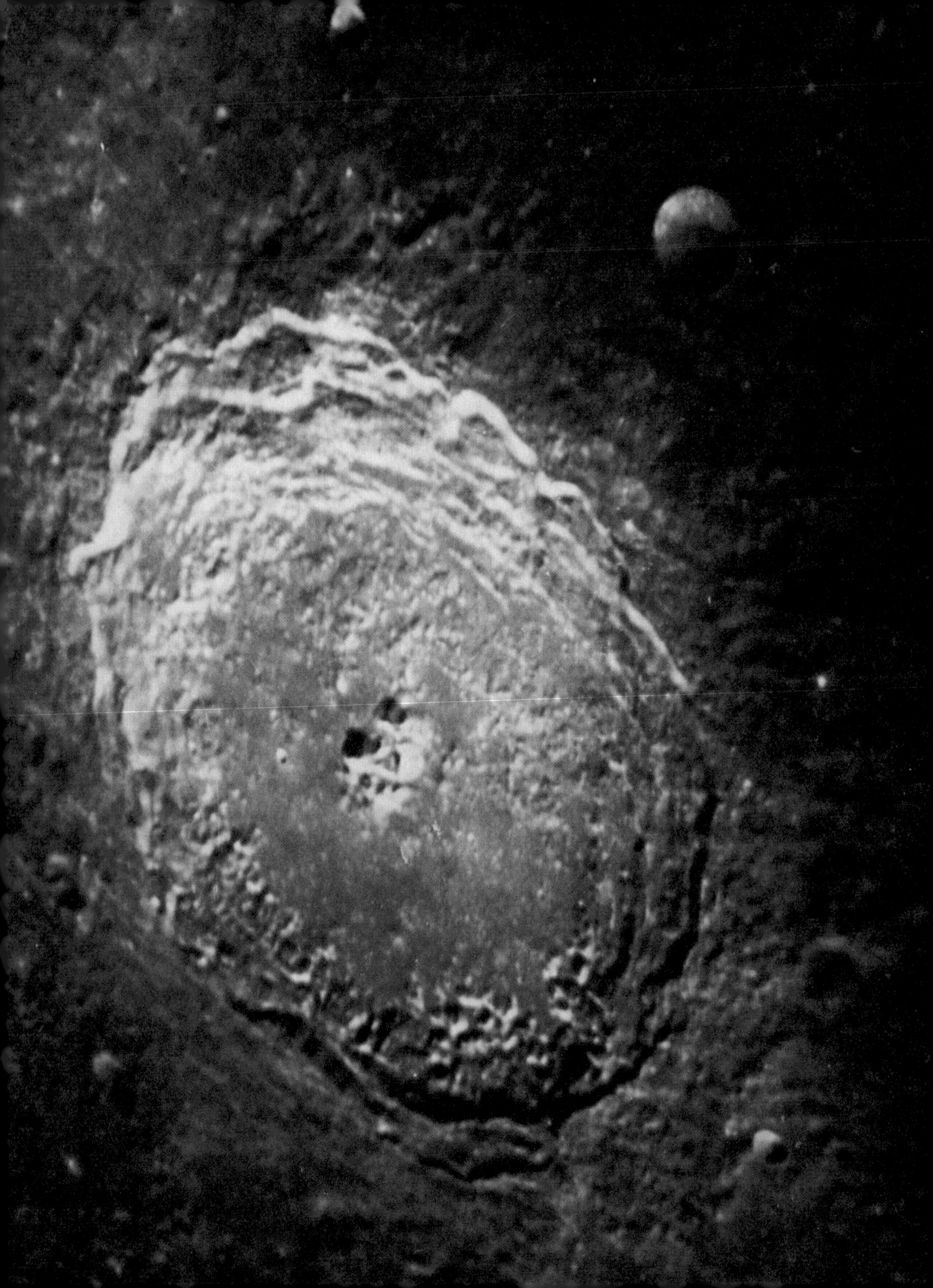

One feature that is quite noticeable through the telescope are the bright rays that come from several of the large craters. These are streaks of rubble that were thrown outward when huge *meteorites*, or rocks from space, hit the moon and formed the craters. The rubble consists of very small pebbles and dust. It does not cast any shadows, and therefore the rays are not seen at other phases of the moon, when the sunlight hits the surface at a sharper angle. Only when the sunlight is shining directly down upon this rubble does it reflect back its almost white color in the form of brilliant rays.

To see the craters and other features more clearly, it is best to view them at the crescent-, half-, or gibbous-moon phases. At these times dark shadows sharply outline the many rough features. The larger the telescope, the more detailed the picture will be. Even before any spacecraft were sent to the moon, astronomers had taken excellent photographs of the lunar surface with large telescopes. Some 30,000 craters could be detected, ranging from less than ⅔ of a mile (1 km) to about 150 miles (240 km) in diameter. However, because of the great distances involved, they could not hope to detect the smaller features it was necessary to see in order to ensure safe lunar landings. What was needed were much closer and more detailed views of the surface, showing features that were too small to be seen from the earth. This could only be accomplished by sending spacecraft to the moon.

A typical moon crater

CHAPTER SEVEN
THE CRATERS– A CLOSER LOOK

Today, after twenty-five years of sending both manned and unmanned spacecraft to the moon, astronomers have a much better picture of our nearest neighbor. Some have even called it the "new moon" because it is so different from our concept of the moon before the space explorations. Let us look closely at this "new moon."

As you can see from the photographs, there are craters of various sizes all over the surface of the moon. They were mainly formed by meteorites crashing into the lunar surface, although some are believed to have been formed by volcanic activity. All of them were created many millions or billions of years ago. The craters with rays coming out from them, such as Tycho and Copernicus, are the youngest ones, but even they are 200 and 600 million years old respectively. Only very tiny craters are being formed today, too small to have any noticeable effect on the appearance of the moon. The moon is basically as it was 3 billion years ago.

By convention, all the craters have been named after famous people, mainly scientists and philosophers such as Tycho, Plato, Copernicus, Kepler, and Aristotle. Because most of the near-side craters were discovered in the seventeenth century, they bear the names of older or ancient scientists and philosophers. The craters discovered more recently generally have more modern names.

In 1959, a Soviet unmanned spacecraft photographed the far side of the moon. This provided a new opportunity to honor famous scientists. The larger far-side craters were named immediately by the Russians, but the rest were just given numbers.

Then, in 1970, when the far side had been more thoroughly photographed and mapped, a special international conference was held to decide upon names for these numbered craters. And still later, as a result of the *Apollo* missions, many of the remaining prominent smaller craters on both sides of the moon were also named, some of them by the astronauts themselves.

The largest crater is Bailley, which is about 180 miles (290 km) in diameter. It is difficult to see from the earth because it is at the edge of the lunar near side. The crater Clavius, which is about 140 miles (225 km) in diameter, is much easier to locate. It is the second largest crater. Most of the others are much smaller. In fact, many are only a fraction of an inch wide. These tiny craters are of course not named or numbered, but they are abundant.

Some areas have only a few craters, while other places seem to be filled with them. In some regions there are so many craters and they are so crowded together that many of them overlap. When a perfectly formed crater lies against the wall of another crater, sometimes breaking down that crater's wall, we can say with great certainty that the perfectly formed crater was created after the one whose wall is broken. This is one of the ways that astronomers were able to date when these craters were formed. In the same way, if one crater is found within another, obviously the inner crater was made later.

Craters are usually circular or oval-shaped, unless they have been distorted due to later bombardments by meteorites. Their original shapes were determined by the size, speed, and angle of the meteorite as it struck the lunar surface. If you throw different-sized stones at various angles and speeds into a bed of wet sand or mud, you can obtain similar results.

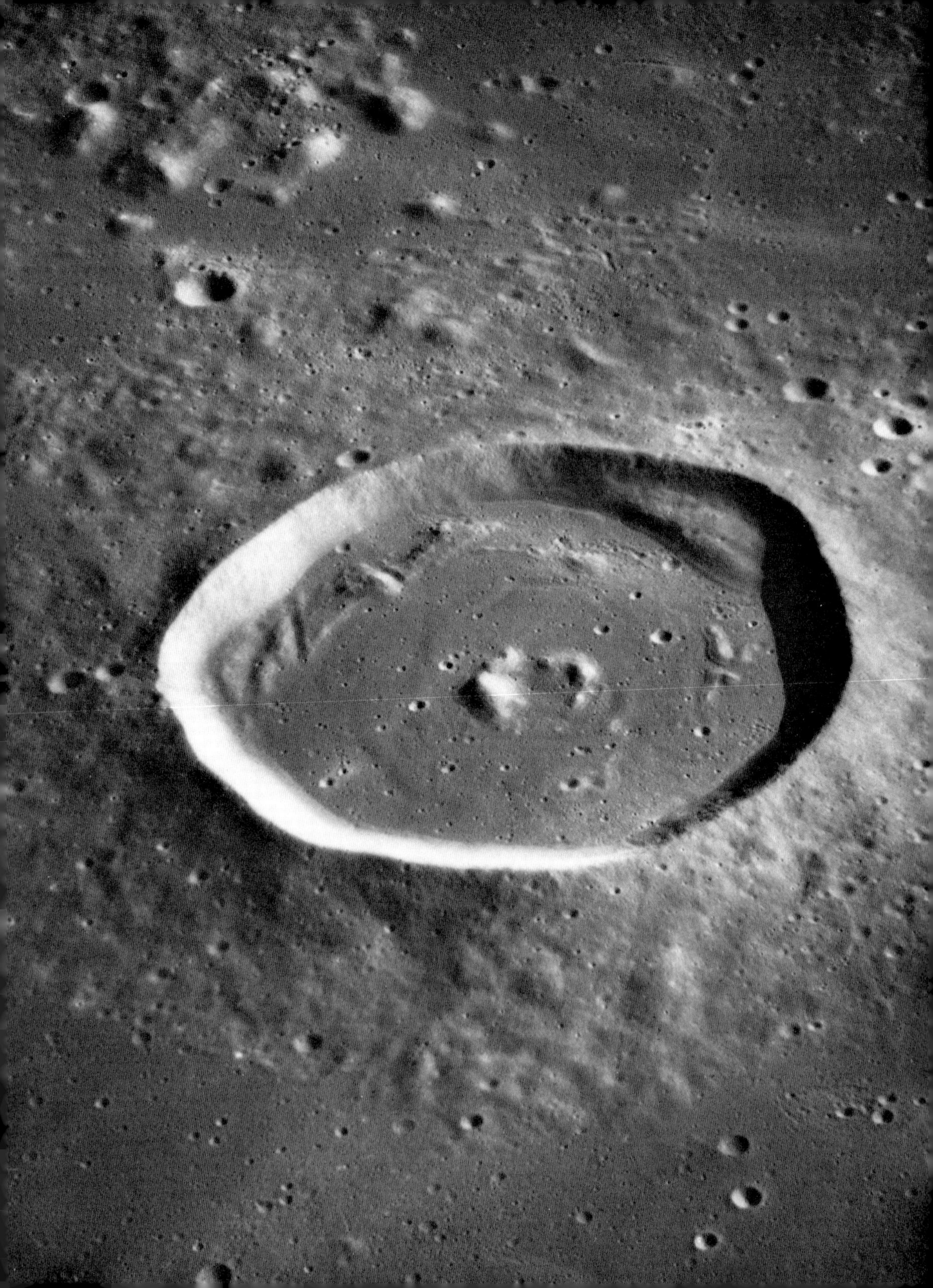

The walls of the craters may reach more than 20,000 feet (6,000 m) above the crater floor. They are much steeper inside the crater than outside, where they usually slope gently down to the surrounding land.

In the center of some of the larger craters you can see mountain peaks. When large meteorites hit the moon, they not only made big holes in the ground, they also generated a lot of heat. This heat was enough to temporarily melt some of the lunar rocks. The molten rock splashed up in the middle of the forming crater, then cooled before it could flow back down onto the crater floor. This formed a peak or mountain in the center of the crater. You may find that some of your sand or mud craters form similar peaks if the rocks are big enough and are hurled down with enough force.

The large crater Kunowsky. Note the many smaller craters on the floor of the large crater and also the central peak.

CHAPTER EIGHT
THE MARIA

In the moon's early formation, very large meteorites, possibly as large as 100 miles (160 km) across and weighing more than 1,000 tons, are believed to have crashed into the newly hardened crust of the lunar surface. These impacts made enormous craters. They also cracked the crust beneath them, making deep fissures extending to depths of 125 miles (200 km) or more.

Beneath this crust was molten lava much like what is in the earth's interior today. This lava oozed up through the deep cracks in the crust and started filling in the deep craters that the giant meteorites had made. This action continued for about 700 million years. The molten lava flowed into the large crater basins, spread out, and hardened, forming what we now call the maria.

By 3.1 billion years ago, the cracks had all been plugged up by hardened lava and the flow stopped. The crust had itself hardened again by then and was very thick. It could not be easily broken apart or shifted about any more. Any volcanic or quake activity on the moon today is very minor and only affects the very deep interior. This is quite different from conditions here on earth, where volcanic eruptions and earthquakes are frequent occurrences.

The maria, then, are lava-filled crater basins. You can still make out the circular shapes of these early craters, although

Kepler
Aristarchus
Oceanus Procellarum
Mare Imbrium
Copernicus
Alps
Apennines
Plato
Mare Serenitatis
Mare Frigoris
Mare Crisium
Mare Fecunditatis
Grimaldi
Gassendi
Mare Humorum
Mare Nubium
Tycho
Ptolemaeus
Mare Tranquillitatis
Mare Nectaris
Vendelinus
Langrenus

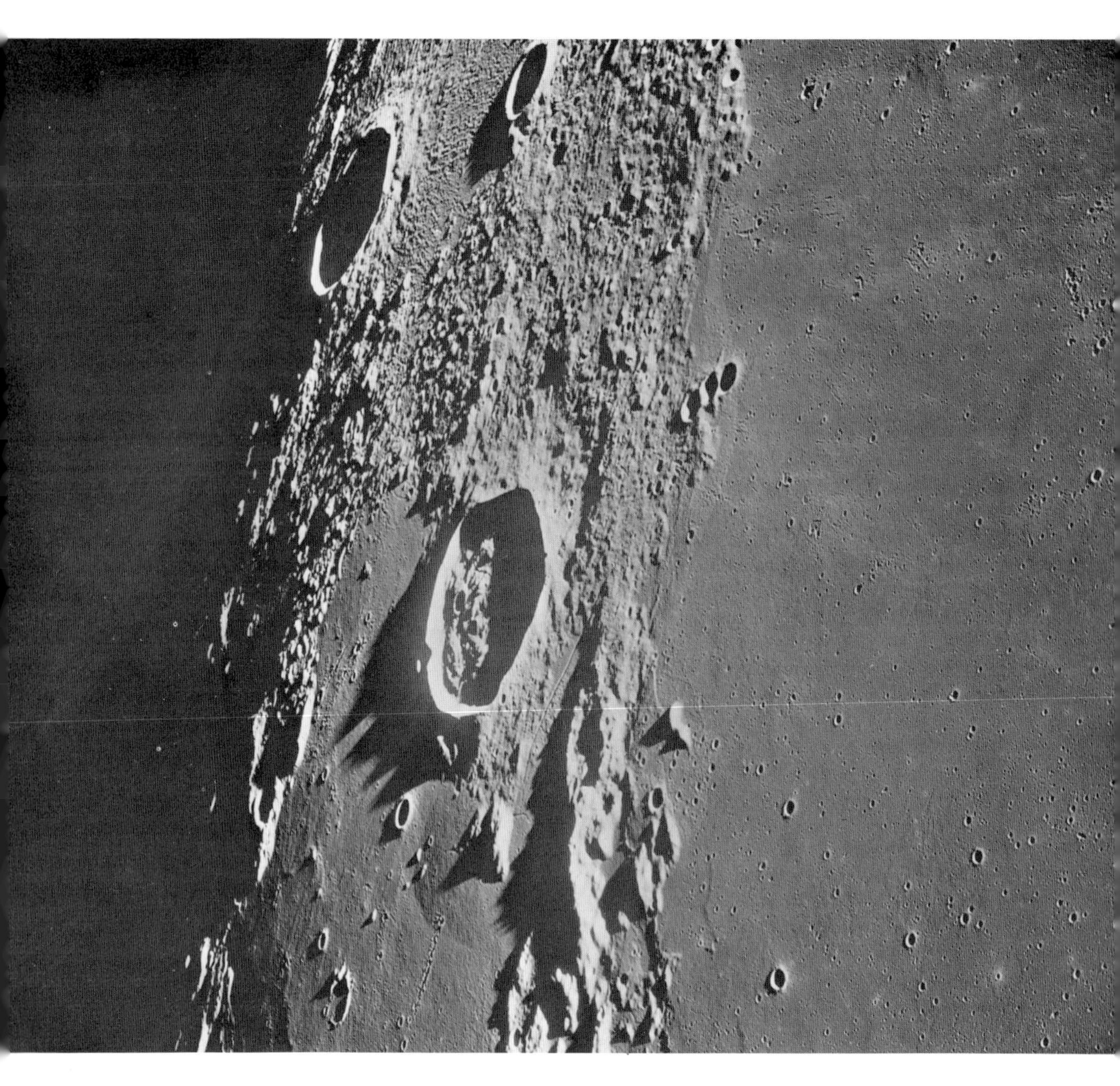

Even the seemingly smooth maria of the moon are covered with many tiny craters.

parts of their walls have been broken down by later meteorite hits. The maria are the largest single feature on the lunar surface. As we have seen, their contrasting darkness against the lighter highlands creates the man-in-the-moon image.

There are more than a dozen large maria. The largest one is Oceanus Procellarum. But this mare is not easily viewed in its entirety from the earth because of its location on the edge of the lunar near side. The largest mare in the central part of the near side is Mare Imbrium (the Sea of Showers). Mare Imbrium has a diameter of about 700 miles (1,120 km). This is quite a bit larger than any of the craters. And unlike the craters, the maria were all given very peaceful, soothing names (Tranquillity, Serenity, etc.). All the names are in Latin, however, because that was the language of science in the seventeenth and eighteenth centuries, when these features were named. Before that time, they were regarded merely as dark patches on what was believed to be a perfectly smooth ball.

Although the large meteorites that hit the moon early in its history made large craters on both sides of the moon, only the side facing the earth, the near side, has maria of any great size. For some reason, the far side never experienced any lava flow. This may have been due to the earth's gravity pulling the lava toward it. Huge, ancient craters made by giant meteorites can still be detected on the far side, but they have been battered out of recognition.

Close-up photographs of the surface, such as those taken by the *Apollo* astronauts, show that the maria are not completely smooth. Their lava is pockmarked with crater holes. Although these vary in size, none is very big, especially compared to the craters that are seen in other areas. The holes in the lava beds were made by small meteorites hitting the surface after the lava had hardened. By that time, there were very few really big meteorites left that could make a crater the size of those made earlier.

CHAPTER NINE
OTHER SURFACE FEATURES: MOUNTAINS, RILLES, AND DUST

Although the maria and the craters are what we think of first when describing the moon, there are many other surface features. For example, the moon also has a few mountain ranges, similar to those on earth. Some of these reach heights of 25,000 feet (7,500 m) or more. This is comparable to our highest mountains. The lunar mountain ranges were named after mountain ranges on the earth. There are the Alps, the Apennines, the Carpathians, and the Pyrenees on the moon as well as on the earth. These higher areas, along with the lower hills, are usually referred to as the *highlands.* This distinguishes them from the maria. The highlands are the brighter areas of the lunar surface.

There are also many canyonlike crevasses and valleys running along the lunar surface. These are called *rilles*. They are either straight lines or they twist and turn, much like riverbeds here on earth. The straight rilles are usually more than 3 miles (4.8 km) wide and hundreds of miles long. They cut across craters and maria and do not seem to be related in any way to any other surface feature. It is thought that they may be similar to fault lines here on earth. Fault lines on earth are breaks in the surface where earthquakes have divided the terrain. Although there is no such activity along these straight rilles on the moon today, the rilles may have been formed by

A view from above of the Hadley-Apennine Mountain region. Hadley Rille meanders through the center of the photo.

An Apollo 15 *astronaut with the Lunar Rover at the edge of Hadley Rille.*

quakes that occurred long before the lunar crust became so rigid. Features on the moon remain for millions of years, unlike those on the earth, which are erased by erosion. There is no erosion on the moon because there is no air or water to cause this gradual wearing away of surface features.

The twisting, or sinuous, rilles have been closely studied, mainly because of their strong resemblance to river channels here on earth. However, detailed examinations and comparisons have shown that the similarity is in appearance only. They are actually more like the formations created by lava flowing out of a volcano here on earth. We call such formations *lava tubes*. It is more likely that the sinuous rilles formed from flowing lava rather than water, since there is absolutely no evidence of any water on the moon, whereas there is plenty of hardened lava.

One sinuous rille was explored by astronauts during the *Apollo 15* mission. This was the Hadley Rille, an 85-mile (136-km)-long channel. It averages about ¾ of a mile (1.2 km) in width and 1,230 feet (1,970 km) in depth. Nearby is the Hadley Delta Mountain, which rises about 13,000 feet (3,900 m) above the plain.

Covering just about everything on the lunar surface is a layer of fine, powderlike soil or dust. In some places it is mixed with rocks and pebbles. It varies in depth from place to place, and when it is mixed with the rocky rubble, it can be as deep as 65 feet (19.5 m). The fine soil is actually the result of billions of years of meteorites hitting the lunar surface and breaking up the rocks. It is ground up, or pulverized, rock. The astronauts were not hampered by its presence on the surface since they never sank more than a few inches into it. However, it did cling to their space suits as they moved about on the lunar surface.

CHAPTER TEN
MOON ROCKS

The surface of the moon is strewn with rocks of all sizes and shapes. The *Apollo* astronauts brought some of these rocks back with them when they returned from the moon. In all, some 843 pounds (380 kg) of moon rocks were brought back to the earth. Each rock was carefully labeled by the astronauts and placed in a separate bag so that scientists back on earth would know where each piece came from. By studying these rocks, much has already been discovered, but even now, more than ten years after the last *Apollo* mission, the data is still being examined and analyzed.

Moon rocks are very similar to rocks found on the earth. Therefore, the names used by geologists for earth rocks were used for the lunar ones as well. Most of the moon rocks were found to be *igneous*, or "fire-formed." Such rocks were once molten and did not harden until after they had cooled off.

Molten lava that cools fairly quickly forms a type of igneous rock called *basalt*. Basalt is very dark in color and has many small holes in it, much like a sponge. A great amount of lava flowed into the very large craters on the near side of the moon, forming the maria. Basalt rocks were found in the maria regions. Here on earth, basalt rocks can be found in the Hawaiian Islands, where there are large lava beds covering much of the land.

Other kinds of igneous rocks formed below the lunar surface rather than on top. It took these rocks much longer to cool, and in the process they grew large crystals that gave them a much lighter color than the dark basalt rocks.

When the lunar surface was hit repeatedly by large meteorites, these rocks were blasted from their original sites inside the moon. They landed on the surface, where they were exposed to even more meteoritic bombardment. The constant pounding upon these rocks broke them into many little bits and pieces and scattered them all over the moon. Some of the meteorite hits were so powerful that they generated enough heat to make some of these rocks molten again. In such a state, many pieces of different kinds of rock got cemented or glued together by the molten mixture and cooled off as one piece. These jumbled rock mixtures are called *breccias,* which means "broken rocks." In them you can see the bits and pieces of the older rocks stuck together.

Breccias are the major kind of rock found in lunar areas not covered by lava. These are the so-called highlands. The highlands are believed to be the oldest lunar surfaces. The highland breccias were shattered and reshattered many times during the billions of years since the lunar crust was formed. Some of the breccia became lunar dust. The basalt rock from the lava flows formed at a much later date.

It is very difficult to recreate the chain of events in the moon's history from such a jumble, but geologists using radioactive dating procedures are managing to obtain a fairly clear picture of the development of the moon.

Among the most spectacular of the materials found in the rocks brought back from the moon were tiny spherical pieces of glass. Most of these were caused by the melting of rocks during the meteorite impacts. Bits of molten rock were splashed away and formed into round balls or sometimes into teardrop shapes. Some have irregular shapes because they hit another object before they hardened. Like the other highland

0 1 2 3 4 5
N1

lunar rocks, some were later shattered by meteorites and became part of breccias.

Around the edges of the maria many bright emerald-green and orange glass balls were found mixed in with the soil or dust. Although these are tiny objects, like the glass balls found in the highlands, there are enough of them to give the soil a definite color in places. It was the unusual color of the soil that alerted astronauts to its different nature. Most of the soil or dust and rocks on the moon are grayish-brown. These colored balls of glass were formed from lava fountains splashing against the edges of the craters before cooling. The tiny droplets contain a large percentage of certain elements that give them their bright and distinctive color.

Moon rocks. The ones above are breccias, heavily coated with glass chips. The one below is a basalt rock.

CHAPTER ELEVEN
THE LUNAR INTERIOR

When the *Apollo* astronauts were on the moon, they placed a series of seismographs in different locations. These instruments are used to detect earthquakes and tremors here on the earth. Of course, on the moon they would be detecting moonquakes and moon tremors. The seismographs placed on the moon were so sensitive that they could detect the astronauts' footsteps as they walked around on the lunar surface. All of the seismic measurements recorded were transmitted by radio signals directly to the earth, where they could be recorded and analyzed.

When a quake occurs on either the moon or the earth, its vibrations go out in all directions. They bounce off different parts of the interior, creating various patterns as they hit the different parts. By studying these patterns astronomers can get a good idea of what the inside of the moon is like.

The outer layer, or *crust*, of the moon is about 37.5 miles (60 km) deep. The top half-mile (.8 km) or so is composed of loosely packed rubble rather than hard rock. The rocks become more and more tightly packed as the depth increases. The crust is somewhat thinner on the near side because of the maria, which push down upon the crust, compacting it even further.

Below the crust is the *mantle*. This layer is much thicker than the crust, extending several hundreds of miles deep. We don't know very much about this part of the moon, although this is where the moonquakes originate. Even though the quakes on the moon are much weaker than those on the earth, their occurrence does mean that there is some activity going on deep inside the moon. It is not completely dead or unchanging.

The innermost part of the mantle is believed to be partly molten, which may contribute to the moonquake activity. Beneath the mantle is a small *core*, which is thought to be made up mostly of iron with a mixture of other minerals. This core may also be partly molten.

When spacecraft first started circling the moon, it was noted that they were pulled slightly out of their orbits whenever they passed over certain areas. These areas were most often in the maria. It was found that the material making up the maria, the hardened lava, is denser than the rocks in the highlands. The lava material is thicker and more concentrated. Because of this, it has a greater gravitational pull on nearby objects. It was this extra force that was pulling the low-orbiting spacecraft off their paths. These areas in the lunar maria were called *mascons* ("mass concentrations"). There are no mascons on the far side of the moon because there are no maria there.

CHAPTER TWELVE
THE HISTORY OF THE MOON

Where did the moon come from? Astronomers have several theories concerning this subject, but so far none can completely explain how the moon came to be what it is today.

Did the moon break off from our rapidly spinning earth to become a separate body? Or did it form elsewhere in the solar system and get captured by the earth at a later date? Did the earth and moon form together as two separate bodies in orbit around each other just as they are found today? Or did some other large object from space collide with the earth in its early formation and blast off a large part of the earth's outer surface? And did the debris from both of these bodies then go into orbit around the remaining earth as separate chunks and bits of matter but later get pulled together by gravity into one mass, that is, the moon?

Any theory must be able to explain why the moon has so much less iron in it than the earth. Why is the moon rich in aluminum and titanium whereas the earth has far more hydrogen, helium, lead, and mercury? The major element found on the moon is oxygen, but most of it is chemically united with silicon, which is the second most abundant element there. There is no free oxygen, the kind used for breathing, on the moon.

If the moon did not originate as a satellite of the earth,

then the theory must explain just how it got there. There are specific laws of physics that put limits on the kinds of motion a body can make while it is in space. These laws govern the earth, the moon, the astronauts and their spacecraft, and so on. They must be considered in any theory about space and motion.

So far, scientists have not solved the problem of where the moon came from. However, from studying the lunar rocks and other information obtained by astronauts, scientists have a pretty clear picture of what happened after the moon was formed.

In its earliest days, the outer layers of the moon (and its interior as well) were very hot and in a molten state. Around 4.5 billion years ago, the moon was probably a bubbling mass of very hot material.

The outer crust gradually cooled and hardened. The type of rock found in the lunar highlands today slowly formed beneath the surface. Then, between 4 and 4.5 billion years ago, most of the huge meteorites came crashing down upon the newly cooled crust and heavily cratered it.

These meteorites were part of the debris being hurled around in the solar system as it was forming. They crashed into all the moons, planets, and asteroids. Astronomers have found evidence of such cratering on all solid bodies in the solar system, including the earth. However, most of the evidence of such heavy cratering on the earth has long since eroded away.

From about 3 to 4 billion years ago, lava from the lunar interior found its way up the deep cracks in the crust to the surface. Flowing into the larger and deeper crater basins, the lava cooled and formed the maria. From then on until the present day, very little change has occurred on the moon. A few new large craters with their distinctive rays have been formed by an occasional meteorite, but basically the moon has remained as we see it today.

CHAPTER THIRTEEN
THE TRIP TO THE MOON

Going to the moon has long been a dream of humankind, but throughout most of history it was considered an impossible goal. In literature, of course, there have been many imaginary journeys. One of the most famous descriptions of such a trip is Jules Verne's *From the Earth to the Moon*. It is a fascinating science fiction story.

The real trips to the moon—the *Apollo* missions—took place in the late 1960s and early 1970s. For the first time, a human being walked upon the surface of a world other than the earth. This was a major turning point in human history and will always stand out in any record of scientific achievement. Commander Neil Armstrong's words upon making that first step onto the lunar surface will long be remembered: "That's one small step for a man, one giant leap for mankind."

To prepare for the *Apollo* missions, astronomers had to have a much better picture of the lunar surface than could be obtained from the earth. Landing sites for the *Apollo* spacecraft had to be selected with great care. Therefore, in the 1960s, before any astronauts were sent to the moon, many unmanned spacecraft were sent there by both the United States and the Soviet Union. These craft were equipped with television cameras to take close-up pictures of the lunar surface and transmit them back to earth. Pictures were taken right up to the

moment of landing or crashing, showing very detailed views of the lunar surface. Now, the first highly detailed mapping of the lunar surface was possible; relatively smooth landing sites could be selected for the *Apollo* missions.

There were six *Apollo* missions that landed men on the moon. Each was made up of a team of three very highly trained astronauts. The trip to the moon took about four days.

What was it like to travel on such a journey?

After blasting off from the earth with the aid of a huge 363-foot (109-m)-tall *Saturn V* rocket, the space capsule went first into orbit around the earth. From there, other rockets were fired, sending the spacecraft on to the moon. The rockets were only used to control the speed and change the direction of the spacecraft. The gravity of the earth and the moon kept the spacecraft moving most of the time.

The rockets sent the spacecraft from earth orbit at more than 35,000 feet (10,500 km) per second (about 24,000 miles, or 38,400 km, per hour), but the earth's powerful gravity slowed the ship down as it moved away. It went slower and slower, coasting toward the moon until it got quite close. When it was about 34,000 miles (54,400 km) from the moon, it was going only 2,900 feet (870 m) per second. Then the moon's gravity took over and pulled the craft down to the lunar surface. As it fell toward the moon, it went faster and faster, like any falling object would. The astronauts now fired their rockets to slow their ship down and keep it from crashing into the moon.

As it neared its destination, the *Apollo* spacecraft was put into orbit around the moon, again with the help of rockets. Two of the three astronauts then climbed into a specially designed lunar landing craft, separated themselves from the command module, and slowly descended to the lunar surface. The lunar landers had four long legs and looked somewhat like giant bugs. In order to get down to the lunar surface once they had landed, the astronauts had to climb down a ladder on the side of one of the legs.

The Apollo 15 *command module in lunar orbit.*

Buzz Aldrin descends to the lunar surface in the first manned landing on the moon.

The third member of the *Apollo* team always stayed in the command module and continued to orbit the moon. This gave him time to take photographs of the moon, earth, stars, and sun. These photographs turned out to be of much higher quality than those that had previously been obtained via television from the unmanned spacecraft.

Many photographs of the far side of the moon were taken while the astronauts were in orbit around the moon. However, while their spacecraft was passing over the far side, they had no contact with the earth at all. Here on the earth, radio waves, which move only in straight lines, can be bounced off different layers of our atmosphere, enabling us to receive "short-wave" radio signals from around the world. On the moon this is not possible. Because there is no atmosphere on the moon, radio signals cannot be sent around to the near side and then to the earth. The lone astronaut left to orbit the moon while his companions explored the surface could neither see nor hear anything from any other human being while he was on the far side.

Once the astronauts on the lunar surface were ready to leave, they climbed back into the small cabin of their lunar lander. The lander then separated from its lower leg supports and blasted off to join the command module in orbit. Careful maneuvering was necessary to hook up the small cabin with the command module. The lower part of the lander stayed on the moon. Once all three astronauts were again together in the command module, the cabin was jettisoned, and it crashed back down onto the moon. Astronomers back on earth were able to detect the crash, thanks to the sensitive seismographs left on the moon.

Rockets were again fired to return the command module with its three astronauts back to earth. Once again, the gravity of the earth and moon were responsible for the motion of the spacecraft. The *Apollo* crews all landed in the Pacific Ocean and were quickly picked up and taken by helicopter to nearby aircraft carriers.

CHAPTER FOURTEEN

MEN ON THE MOON

During each *Apollo* mission, when the two astronauts on the moon wanted to leave the lunar lander and explore the area around them, they first had to put on bulky space suits. These protected them from the intense heat of the sun as well as from other forms of dangerous radiation. Here on earth, our atmosphere keeps most of this radiation from reaching us. The moon has no atmosphere and is therefore completely exposed.

Because there is no air on the moon, the astronauts also had to carry oxygen tanks. On the earth, all of this extra equipment would weigh about 84 pounds (37.8 kg). It would be difficult to walk anywhere with such a load. But the moon is a much smaller body than the earth and therefore does not exert as much gravitational pull on objects on its surface. Everything weighs much less on the moon. If you weighed 100 pounds (45 kg) here on the earth, you would weigh only about 16 pounds (7.2 kg) on the moon. The portable life-support packs carried by the astronauts weighed a mere 14 pounds (6.3 kg) on the moon.

Because there is less gravity on the moon, it takes less effort to lift a heavy object or throw it. But it also requires more effort to push a shovel into the ground by stepping on it. Your weight is much less on the moon, and when you use it to push something down, it is not as effective as it is on the earth. When

Buzz Aldrin performs an experiment on the moon's surface. The lunar module is in the background.

they were collecting specimens of moon rocks, the astronauts found it hard to dig into the lunar soil with shovels.

Although their space suits were big and awkward, the astronauts found little difficulty in moving about. They soon learned how to hop around, somewhat like kangaroos, covering many feet with each jump. Once again, the weaker gravity allowed them to move quickly in this manner.

In later *Apollo* missions, a Lunar Rover was brought to the moon in the spacecraft. This was a four-wheeled vehicle weighing about 400 pounds (180 kg) on the earth. With it the astronauts could explore from 3 to 5 miles (4.8 to 8 km) from their landing site, riding in the Rover at up to 10 miles (16 km) per hour. Because there is no air on the moon, a gasoline engine, which requires oxygen to operate, would be useless. The Rover operated on battery-powered electric motors. It had a color television camera mounted on it that could be controlled by scientists on earth. In this way the many activities of the astronauts could be watched.

In order to communicate with each other when they were in their space suits, the astronauts had to use radio intercoms. Sound waves do not travel through empty space; they require air or water or some other medium through which to move. There is no air or water on the moon, and therefore there is no sound. Radio waves, like light waves, can move through empty space and can therefore be used for communication on the moon. Of course, when the astronauts went back inside the cabin of the lunar lander, they could take off their space suits and talk normally. The cabin was filled with air.

The astronauts kept earth time during their stays on the moon. However, even though several of our days had passed, the sun never set for them. It moved through only a small arc in the sky. They would have had to remain on the moon for a week or more to experience a sunset, because a "day" on the moon lasts two earth weeks. After that comes two weeks of night.

When the sun finally does set on the moon, darkness

comes immediately. There is no gradual twilight, such as we are accustomed to here on earth. Our atmosphere scatters the sunlight even after the sun has gone below the horizon, so that we still have some light. This does not happen on the moon. The sun also rises abruptly on the moon, illuminating the rocky landscape. But the sky remains black because there is no atmosphere to scatter the blue rays of the sunlight and make the sky appear blue.

As the lunar day progresses, the surface gets hotter and hotter. By midday, when the sun is directly overhead, the lunar surface is a boiling 265° F (129° C). No wonder the astronauts needed protective visors for their eyes and very thick space suits.

The night temperature on the moon is no more pleasant. By the middle of the lunar night, the temperature has dropped to a deep freeze of around -310° F (-190° C). Of course, the astronauts never experienced this cold because they were never on the night side of the moon. The *Apollo* missions were planned for daytime exploration only.

Since they landed on the near side of the moon, the earth was always visible to the astronauts. It did not move from its place high in the dark sky. It shone with a brilliance of almost fifty times that of the full moon on the earth. It was important that the astronauts land on the side of the moon facing the earth so that radio contact with the earth could be constantly maintained.

The astronauts described their trips on the lunar surface in terms of going north or south, just as we would do here on the earth. However, we know which way is north because our compass needle points in that direction. The needle is pointed north by the earth's magnetic field. The moon does not have a magnetic field, although there is some evidence that one may have existed in the distant past. A compass will not work on the moon. How could the astronauts tell which was north or south?

Like the earth, the moon rotates on its axis. The earth

turns once every 24 hours; the moon turns once a month (by earth timekeeping). The North and South Poles on the earth are the two points through which the earth's axis passes. The moon has two similar poles, and therefore scientists agreed that the lunar pole oriented most nearly like the North Pole of the earth would be the north pole of the moon.

There is no air or water on the moon, and therefore there is no weather. There is no rain or snow or wind. There are no clouds. On the earth, these are the elements largely responsible for changing our landscape. Wind and rain and flowing rivers are constantly wearing away some features on the earth and building up others.

These changes do not occur on the moon. As far as we know, the craters and mountains and lava beds have been as we see them today for millions of years. The only changes in recent times came from the visiting astronauts. They left their footprints in the dusty surface. They left many instruments and parts of their spacecraft behind. The Lunar Rover was left there also. Nothing will happen to these things unless some future astronauts disturb them. Like the craters and the rocks, they will remain unchanged, possibly for another billion years.

CHAPTER FIFTEEN
WHAT NEXT?

The scientific aims of the *Apollo* missions were to obtain measurements and data that could not possibly be obtained from the earth or by unmanned spacecraft. It was a most successful series of missions. The moon became a kind of scientific laboratory; many types of instruments were brought there, and some were left as semipermanent installations. From the great amount of information collected, many of the mysteries of our nearest neighbor are beginning to be solved. And, as is usual with such scientific investigations, other questions have arisen that were never even thought of before. Much research is still being conducted based upon the data brought back by the astronauts.

We have already seen how valuable the seismographs have been in obtaining information about the lunar interior and the history of the moon. The same is true of the 843 pounds (380 kg) of moon rocks that the astronauts brought back with them.

Special laser-reflecting mirrors that were left on the moon have been used to determine with great accuracy the earth-moon distance. They also can help geologists determine how much and how fast the continents here on earth are moving relative to each other. Studies of "continental drift" are a very big part of research concerning the earth.

Thermometers left on the moon have recorded temperature changes on the lunar surface and also the flow of heat between the lunar interior and its surface. Other instruments have recorded the number of meteorites and atomic particles that are hitting the surface. These data will help scientists plan what kind of protection is necessary for any prolonged future mission to the moon.

The moon certainly doesn't sound like a place where you might want to live or even go to for a vacation. If lunar colonies are ever built on the moon, as has been envisioned by many science fiction writers, they will have to be domed and insulated. The people living there will have to import air to breathe, water to drink, and food to eat; they will have to have protection from the extremes of temperature, as well as from the deadly radiation coming from the sun and outer space.

At present, however, astronomers are thinking more in terms of a possible moon-based observatory. Such an installation would have many important advantages over telescopes in earth-orbiting satellites, although both would make atmosphere-free observation possible. The moon provides a stable platform and allows continuous observations for periods of fourteen days (earth days, that is). A lunar telescope, particularly if placed on the far side, would be free of all earth glare and other terrestrial interference. A radio telescope on the far side would be free of man-made electrical interference, something that is at times a great problem to radio astronomers here on earth. Not only astronomers, but physicists, chemists, biologists, and other scientists who greatly need a "vacuum" laboratory free from atmospheric and electronic interference would find the moon a good place on which to set it up. The moon must certainly be considered in planning any future space ventures.

GLOSSARY

Annular eclipse—a solar eclipse in which only the center part of the sun is covered by the moon because the moon is too far from the earth for its shadow to reach the earth's surface.

Basalt—igneous rock that was once molten lava.

Breccias—rocks that are mixtures of all kinds of rocky material.

Command module—that part of the *Apollo* spacecraft that stayed in orbit around the moon and brought the astronauts back to the earth.

Corona—the outermost part of the sun's atmosphere; it is only visible during a total solar eclipse.

Crater—a hole or depression made in the surface of a moon or planet by a meteorite crashing into it.

Crescent moon—a lunar phase in which only a small part of the moon is visible.

Crust—the outer layer of the moon.

Diamond Ring effect—the effect produced by the last rays of the sun shining between the lunar mountains immediately before or after totality is reached during a total solar eclipse.

Eclipse—the cutting off of the light of one body by another body passing in front of it.

Far side—the lunar hemisphere that always is turned away from the earth.

Full moon—a lunar phase in which the entire near side of the moon is visible.

Gibbous moon—a lunar phase in which the moon appears larger than a semicircle but not as big as a full circle.

Half-moon—a lunar phase in which the moon appears as a semicircle.

Harvest moon—the full moon that occurs nearest to the beginning of autumn.

Igneous rock—rock that was once hot and liquid and has now hardened and cooled.

Lava—hot liquid rock.

Lunar—having to do with the moon.

Lunar eclipse—an eclipse of the moon. This occurs when the moon passes through the earth's shadow.

Lunar lander—that part of the *Apollo* spacecraft that descended to the lunar surface.

Lunar Rover—a vehicle used on the moon by the *Apollo* astronauts to make exploration easier.

Mantle—the layer of lunar material that lies below the crust, or outer layer, of the moon.

Maria—(sing., *mare*) huge lava-filled craters on the moon once believed to be seas.

Mascon—areas on the moon that have denser or thicker material in them and therefore greater gravitational attraction than the rest of the moon.

Meteorite—rocky debris from outer space that hits the surface of a moon or planet.

Moon—a natural satellite of a planet; the earth's natural satellite is called "the moon."

Neap tide—the lowest high tide of the month.

Near side—the lunar hemisphere that always faces the earth.

New moon—a lunar phase in which none of the near side of the moon is visible.

Partial eclipse—a solar eclipse in which only a portion of the sun is covered by the moon.

Phases—the different apparent shapes of the moon as it revolves around the earth.

Rille—a canyonlike crevasse or valley found on the lunar surface.

Solar eclipse—an eclipse of the sun. This occurs when the moon passes between the sun and the earth.

Spring tide—the highest tide of the month.

Tides—the effect of the gravitational pull of the moon and sun upon the earth, especially upon the earth's large bodies of water.

Zodiac—the twelve constellations through which the sun, moon, and all the planets appear to move.

FOR FURTHER READING

Branley, Franklyn M. *Pieces of Another World: The Story of Moon Rocks.* New York: Thomas Y. Crowell Co., 1972.

Kopal, Zdeněk. *A New Photographic Atlas of the Moon.* New York: Taplinger Publishing Co., 1971. (excellent photographs with descriptions)

Lewis, Richard C. *Appointment on the Moon.* New York: Ballantine, 1969. (describes early moon shots and first *Apollo* landing in detail)

Taylor, G. Jeffrey. *A Close Look at the Moon.* New York: Dodd, Mead, 1980.

Zim, Herbert S. *The New Moon.* New York: William Morrow and Co., 1980.

You might also consider subscribing to a magazine. *Odyssey* is an excellent astronomy magazine for young readers. *Astronomy* is another excellent magazine which is beautifully illustrated. Both can be ordered from 625 E. St. Paul Ave., P.O. Box 92788, Milwaukee, Wisconsin 53202.

INDEX